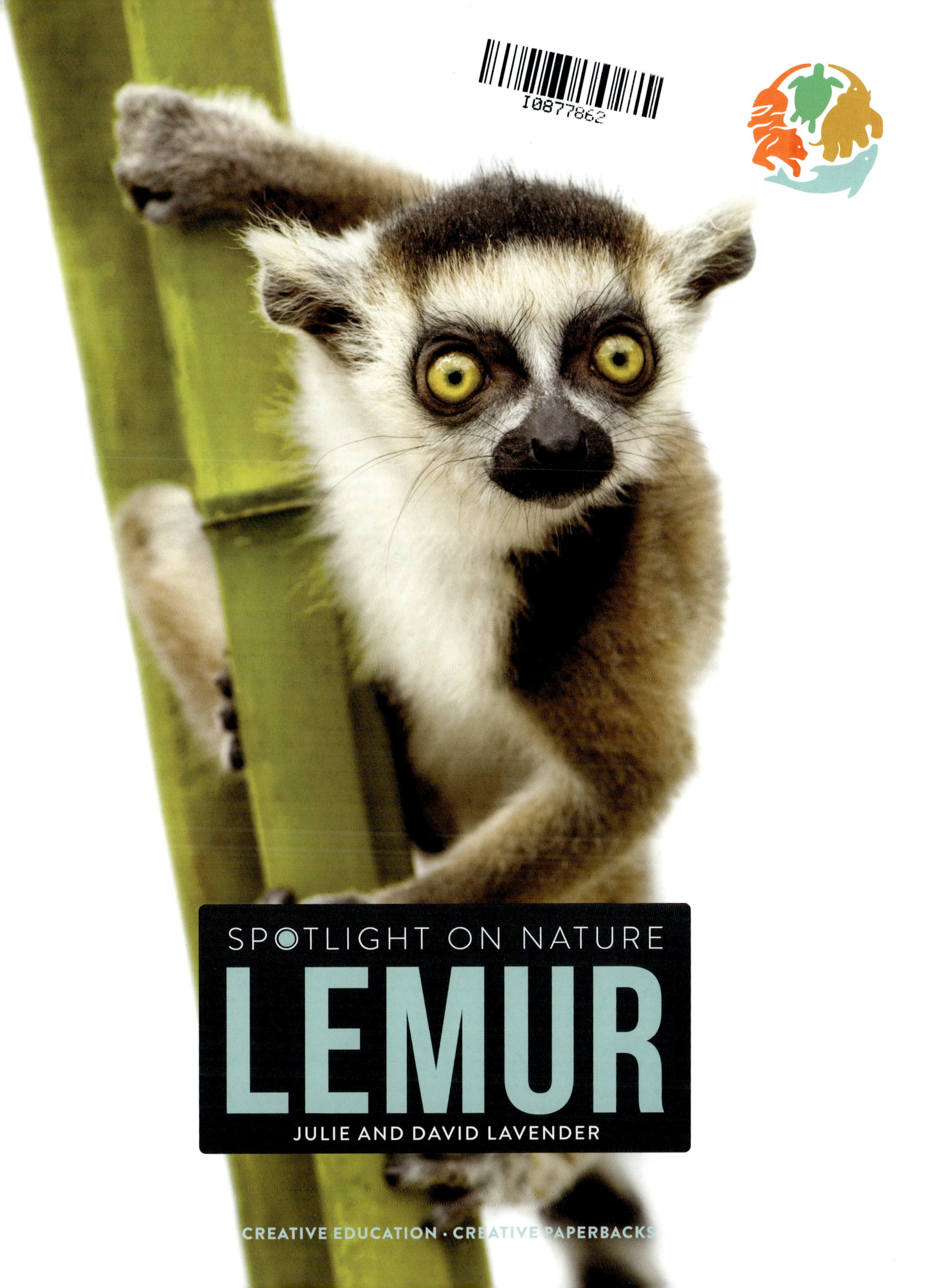

SPOTLIGHT ON NATURE

LEMUR

JULIE AND DAVID LAVENDER

CREATIVE EDUCATION • CREATIVE PAPERBACKS

Published by Creative Education and Creative Paperbacks
P.O. Box 227, Mankato, Minnesota 56002
Creative Education and Creative Paperbacks are imprints
of The Creative Company
www.thecreativecompany.us

Design by Jen Bowers; production Beeline Media & Design, Inc.
Art direction by Tom Morgan
Printed in the United States of America

Photographs by Getty Images (Thorsten Negro, Yann Guichaoua-Photos, ROMAOSLO, AYImages, G & M Therin Weise), Shutterstock (Eric Isselee, Adhivaswut, Jearu, Ondrej Prosicky, Jolanda Aalbers, Port_Folio, Hajakely, sasha_gerasimov, Seregraff, Hajakely, The Hidden Eye, Vaclav Sebek, Ondrej Prosicky, Sander Groffen, RMMPPhotography, Dudarev Mikhail, Ondrej Prosicky)

Library of Congress Cataloging-in-Publication Data
Names: Lavender, Julie, author. | Lavender, David R., 1961- author.
Title: Lemur / Julie and David Lavender.
Description: Mankato, Minnesota : Creative Education and Creative Paperbacks, [2025] | Series: Spotlight on nature | Includes bibliographical references and index. | Audience: Ages 10-13 | Audience: Grades 4-6 | Summary: "Leap into the lemur's world with our new book for mid-level readers. An engaging narrative of a lemur family, captivating images, infographics, milestones, and resources make it perfect for young nature enthusiasts"-- Provided by publisher.
Identifiers: LCCN 2023059463 (print) | LCCN 2023059464 (ebook) | ISBN 9798889891949 (library binding) | ISBN 9781682775677 (paper-back) | ISBN 9798889892021 (ebook)
Subjects: LCSH: Ring-tailed lemur--Juvenile literature. | Lemur (Genus)--Juvenile literature. | Ring-tailed lemur--Life cycles--Juvenile literature. | Ring-tailed lemur--Behavior--Juvenile literature. | Ring-tailed lemur--Madagascar--Juvenile literature.
Classification: LCC QL737.P95 L38 2025 (print) | LCC QL737.P95 (eb-ook) |
DDC 599.8/3--dc23/eng/20240125
LC record available at https://lccn.loc.gov/2023059463
LC ebook record available at https://lccn.loc.gov/2023059464

CONTENTS

MEET THE FAMILY

RING-TAILED LEMURS

of Berenty Reserve

Near the southern tip of the island of Madagascar, ancient tamarind trees form a canopy over the Mandrare River. Farther from the river, dry scrub with tall, thorn-covered plants fills the landscape. Tourists, scientists, and researchers come to this region to visit a small wildlife preserve, Berenty Reserve. At least half a dozen lemur species make their homes here. Other animals in the preserve include Madagascan flying foxes, sickle-billed vangas, knob-billed ducks, Madagascar tree boas and ground boas, spider tortoises, mongooses, and fossas.

On a hot, humid afternoon near the beginning of the rainy season, animals seek shade in the boughs of a tamarind tree. On one branch, a warty chameleon flicks its tongue to catch an insect. Nearby, a female ring-tailed lemur rests on a limb. She has moved away from her troop because she knows something exciting is about to happen. For 141 days, an infant has been developing inside her. Now, it's time to give birth.

CLOSE-UP

A wet nose

Lemurs are classified as wet-nosed primates. The hairless area around the nose stays moist, increasing the lemur's ability to smell. This characteristic separates lemurs from other primates like monkeys and humans.

CHAPTER ONE

LIFE BEGINS

Lemurs are **primates** that live only on Madagascar, the fourth largest island in the world, and the nearby Comoro Islands. There are over 100 known species of lemurs, but this number changes often as scientists identify new types of lemurs. Most lemurs have large eyes, pointed snouts, and long tails. They have woolly fur in shades of gray, brown, black, red, and white. A lemur's extra-long, furry tail helps it balance. Five **digits** on each hand and foot help a lemur climb and leap from tree to tree. Lemurs have keen night vision and an excellent sense of smell.

The smallest lemur, the mouse lemur, weighs just one ounce (28 grams). Its body is four inches (10 centimeters) long, and its tail is even longer at five inches (13 cm). The largest lemur, the indri lemur, weighs 22 pounds (10 kilograms) and is three feet (90 cm) long. Male and female lemurs are generally about the same size.

RING-TAILED LEMUR MILESTONES

DAY 1

- Born
- Weight: less than 3 ounces (85 g)
- Length: 4 inches (10 cm)
- Eyes open
- Has a large head and thin arms

Each lemur species has unique adaptations for surviving in its environment. Lemurs have adapted to live in all the habitats found on Madagascar. These include rainforests, spiny forests, dry forests, rocky cliffs, deserts, plateaus, and coastlines. Different food sources are found in each habitat. Most lemurs are herbivores that eat plant parts including leaves, fruits, flowers, seeds, nectar, gum, and sap. Some lemurs are omnivores that eat both plants and meat. For example, the diet of the omnivorous mouse lemur includes insects, small chameleons, tree frogs, leaves, fruits, and flowers.

CLOSE-UP

Gripping hands

A lemur's tail is not **prehensile**, but its hands are. This grasping ability allows lemurs to firmly grip tree branches. Large pads on the hands and feet provide extra traction when jumping through the trees.

FEATURED FAMILY

Welcome to the World

The female ring-tailed lemur stretches uncomfortably and makes a sharp cry. She rests until her muscles contract again, then stands and grasps the branch overhead. It's dusk now. She crouches in a sitting position and waits. Soon, the infant's head appears. The baby, a male, is born. His eyes are open. The new mother licks him clean. He weighs less than three ounces (85 g) and measures four inches (10 cm) long. His thin arms are scrawny, but his clasping instinct helps him cling tightly to his mother's belly.

Lemurs live in groups called troops. Troops can have as few as three lemurs or as many as 20. One dominant female leads most lemur troops. Each day, she decides what direction the troop will take to forage for food.

Most lemurs give birth to one infant at a time, though some species have up to six babies at once. An infant clings to its mother's belly for two to three weeks and survives on her milk. After that, a young lemur rides on its mother's back until it is fully **weaned**. Once a lemur baby begins piggyback riding, it starts to sample the foods its mother eats. Most lemur males do not help care for their young. Adult females in the troop share parenting duties in an arrangement called alloparenting.

1 WEEK

- Lives on mother's milk
- Clings to mother
- Crawls about on mother's belly

2 WEEKS

- Rides on mother's back
- Grasps at twigs
- Samples some of mother's food

CLOSE-UP

State of dormancy

When food is scarce, a fat-tailed dwarf lemur may enter a state of dormancy called torpor. Inside a hole in a tree, it curls into a tight ball and lowers its body temperature to save energy. For up to six months, it survives on fat stored in its tail.

FEATURED FAMILY

First Meal

The mother lemur pulls her helpless infant to her nipple for nursing. The newborn can't eat solid food yet. At three days old, he can move around on his mother's stomach as he clings tightly to her with his hands and feet. After two weeks, he rides on her back. As he rides, he grasps twigs and samples bites of green leaves. By three weeks of age, the curious lemur begins to hop off his mother's back. He explores branches on his own, but he doesn't stray far.

Young LEMURS ride on their MOTHER'S BACK.

3 WEEKS

- Hops off mother's back for short periods of time
- Samples a greater variety of food

1 MONTH

- Climbs short distances
- Explores surroundings, but doesn't stray far
- Looks for own food to supplement mother's milk

CLOSE-UP

Stink fights

Male ring-tailed lemurs often compete for a female's attention. They rub their long tails against their scent glands, then wave or flick them at their opponent. The stinkiest tail wins!

CHAPTER TWO

EARLY ADVENTURES

By the time they are five months old, young lemurs are weaned and must find food on their own. Most lemurs are **arboreal**. Foraging involves climbing trees to nibble on leaves, flowers, and fruit. Lemurs are especially good at leaping and bounding rapidly from tree to tree. This behavior is known as ricochet **locomotion**. From their position on a tree limb, lemurs push off with their strong back legs. They practically fly through the air to the next tree, leaping as far as 33 feet (10 meters). Their tails act as rudders. To land, their front legs grasp the tree first, and the back legs follow.

Although most lemurs prefer the treetops, many are also quite agile on the ground. Some use quadrupedal locomotion to walk on all four legs. Sifaka lemurs stand upright and walk on their back legs. They bounce and jump in a sideways motion with their arms outstretched for balance. This form of locomotion has earned them the nickname "dancing lemurs."

(2) MONTHS

- Plays with other young lemurs in the troop
- Protected by adult females in the troop
- Follows the troop by tracking their black-tipped tails

CLOSE-UP

Special fingers

Aye-aye lemurs have an extra-long third finger on each hand. These special fingers tap on wood to find hollow tunnels where insects hide. They are also good for reaching inside the tunnels to snag grubs and other bugs.

FEATURED FAMILY

Look Who's Playing

The eight-week-old lemur plays with other young ring-tails. The youngsters jump and wrestle under the watchful eyes of adult females in the troop. When a playmate picks up a stone, the young lemur snatches it and scampers up a tree to hide on a leafy limb. From the ground, the adult females chirp and scold. The tree's leaves rustle and shake, and the young lemur realizes he's not alone. A Madagascar tree boa inches toward him. The young lemur drops the stone and scrambles back to the safety of his troop.

Some lemur species are mostly active at night, and others stay awake during the day. Whether they are nocturnal or diurnal, young lemurs must learn to avoid predators around the clock. Fossas, mammals that resemble small cougars, like to prey on lemurs. They are quick and agile, and they move through trees just as easily as lemurs do. A fossa can pounce on a lemur infant and kill it instantly. Harrier hawks and buzzards can swoop down and snatch young lemurs. Boa constrictors and wild dogs are also predators. Adults in the troop sound warning calls when they spot a fossa or another predator.

5 MONTHS

- Fully weaned
- Forages for food independently during the day
- Ricochets from tree to tree
- Sleeps in a troop huddle at night with tail wrapped around body for warmth

CLOSE-UP

Scratch and scent

Male ring-tailed lemurs have a hard spur, like a thick fingernail, located on their wrist next to a scent gland. They use it to scrape the bark off a tree and then fill the scar with their scent to mark their territory.

FEATURED FAMILY

Give It a Try

Though he's not yet completely weaned, the three-month-old lemur often eats tasty leaves, seeds, flowers, and fruits. Today, the troop makes its way to the river for a drink. The young lemur lags behind the adults, but he keeps a watchful eye on their banded tails lifted high in the air. He approaches the riverbank, searching for young plant shoots. Suddenly, he hears a splash. A Nile crocodile lunges out of the water. Lemurs shriek and scatter back to the forest. The young lemur learns it's safer to stay near the trees.

SIFAKAS are *often* called DANCING LEMURS.

⑥ **MONTHS**

- Uses tooth comb for grooming
- Grooms other young lemurs
- Spends much of the day resting or basking in the sun

⑨ **MONTHS**

- Skilled at finding a variety of foods, including tree resin and sap, spiders, birds, and termites
- Participates in the complex social interactions of the troop

CLOSE-UP

Tooth comb

Most lemurs have special front teeth that are long and slender and that have gaps between them like a comb. They use this "tooth comb" to groom their fur or another lemur's fur.

CHAPTER THREE

LIFE LESSONS

Adult lemurs spend more time resting than they do foraging and feeding. When they aren't eating or sleeping, they might be found grooming themselves or other lemurs in the troop. Grooming cleans the fur and helps it hold warmth. More importantly, it strengthens social bonds within the troop. This **tactile** communication helps lemurs feel close to each other.

Many lemurs spend time basking in the sun for warmth. They do this energy-saving behavior by sitting on their haunches with their backs arched and their faces turned toward the sun. They rest their extended arms on their knees.

Lemurs communicate through sights and sounds. Young lemurs learn communication skills from their troop. They learn how to make facial expressions that show submission and how to react when another lemur gives them a threatening stare. They also learn a wide variety of **vocalizations**. Lemurs use more than a dozen distinct calls and sounds, from shrieks of alarm, to purrs made toward an infant, to early-morning wails that can be heard for miles.

1 YEAR

- Now past the age of greatest threat from predators
- Females help raise younger lemurs
- Males watch older males stink-fight
- Females begin vying for rank within the troop

Lemurs also communicate through scent. Smells provide information about a lemur's health, sex, and territorial boundaries. Larger lemur species tend to be diurnal. Many of them have scent glands on their bodies that are used for scent marking. Smaller lemur species tend to be nocturnal. They often mark their territories with urine or feces.

Lemurs reach sexual maturity when they are two or three years old. By then, they are fully grown, and they've learned the skills they need to survive in the wild. Adult females are ready to care for young lemurs. They are strong enough to carry an infant on their bellies. Or, like the gray mouse

CLOSE-UP

Two tongues

Lemurs have two tongues. Their second tongue, the sublingua, doesn't have taste buds. It is used like a toothbrush to clean the tooth comb.

FEATURED FAMILY

This Is How It's Done

The growing lemur has watched others in the troop forage for fruit and catch insects. He has practiced leaping from branch to branch to search for food. He's ready to scour limbs and twigs on his own to find tasty morsels. He spots two plump caterpillars camouflaged on a limb and snatches them up for a protein-filled snack. Searching onward, he finds an unattended bird nest. He peers inside, finds three eggs, and gobbles them up. He bounds away through the treetops just before the mother bird returns.

Grooming STRENGTHENS SOCIAL BONDS within the troop.

3 YEARS

- Reaches sexual maturity
- Females stay with the troop
- Males may leave to find new troop

4 YEARS

- Females give birth to a single infant each year
- Males may leave to find a new troop every two to five years

lemur, they know how to carry an infant in their mouths before it is old enough to hold onto the mother's back. Three-year-old black-and-white ruffed lemur females know how to build a protective nest for new offspring and suspend it between tree branches.

Female lemurs tend to stay with the troop they were born into. When males become adults, they often leave to join a neighboring troop. Males may belong to several troops during their lives. Most lemurs live for 16 to 20 years. Smaller lemurs, like the mouse lemur or dwarf lemur, have a life expectancy of six to eight years.

FEATURED FAMILY

Practice Makes Perfect

Now five months old and completely weaned, the young ring-tailed lemur receives very little care from his mother. Other troop members groom him, and he grooms them in return. He forages on his own, taking care to stay close to the troop. At night, the troop huddles together for warmth. They form a tight clump with their tails wrapped around their bodies. When morning comes, the young lemur climbs to the tip-top of a tamarind tree for his favorite fruit. He weighs less than the adults, and he can seize the highest fruit that they can't reach.

CLOSE-UP

Toilet claw

The "toilet claw" is a long toenail or claw found on a lemur's second toe. This claw is mostly used for grooming, but it is also useful for scratching an itch.

12 YEARS

- Infants of older mothers have reduced survival rates

19 YEARS

- End of life

CHAPTER FOUR

CRITICALLY ENDANGERED

Lemurs are in trouble. Scientists consider 98 percent of all lemurs to be endangered or critically endangered. That's practically the entire lemur population, making lemurs the most endangered group of mammals.

Many lemur species that once lived are already **extinct**. Members of one ancient lemur species were almost as large as gorillas. For a time, the greater bamboo lemur was thought to be extinct. Scientists found some surviving greater bamboo lemurs, but only about 500 remain in the wild. Other lemur species are on the brink of extinction.

Lemurs are losing their habitats. In the last 60 years, Madagascar has lost almost half of its forests to illegal logging and wildfires that got out of control. Illegally hunting lemurs and collecting lemurs for the pet trade are also devastating lemur populations.

Many people are trying to help lemurs. The Wisconsin National Primate Research Center studies lemurs to learn how to better care for all primates and increase their quality of life. The Duke Lemur Center (DLC) in North Carolina houses over 200 lemurs. Researchers study them and learn how to better protect them. The DLC sends teams to Madagascar to educate local communities about the importance of preserving natural habitats for lemurs. It also plants trees to help reforest the land. The Lemur Conservation Network also works in Madagascar to protect forests and lemurs.

In Florida, the Lemur Conservation Foundation cares for and breeds six species of endangered lemurs. The organization built a base camp in the Madagascar rainforest where researchers can study lemurs in the wild. Zoos all over the world take care of lemurs, breed lemurs, and carry out missions to increase their numbers. World Lemur Day is celebrated every year on the last Friday in October to help educate people about these special primates.

Scientists used to believe that lemurs were less intelligent than monkeys and apes. A study at Duke University asked lemurs to carry out tasks and solve problems using touch screens. They found that lemurs are very intelligent. They are capable of memorizing sequences of pictures and choosing which groups have larger or smaller numbers of items.

Leaping, colorful, playful lemurs are loved and admired by people everywhere. Without our help, they may not survive.

FAMILY ALBUM

SNAPSHOTS

Golden bamboo lemurs eat a type of bamboo that contains toxic cyanide. The amount they eat should be lethal, but they somehow thrive on this diet.

To warm up on cool mornings, **red ruffed lemurs** gather to bask in the sun. They sit upright and face the sun with their arms extended.

Aye-ayes have large front teeth for gnawing. Like a rodent's front teeth, these incisors are continuously growing and self-sharpening.

Blue-eyed black lemurs are the only primates, besides humans, that have blue eyes.

Critically endangered **silky sifaka lemurs** have brilliant white fur. People in Madagascar call this lemur "angel of the forest."

All **ring-tailed lemurs** have exactly 13 black and white bands on their tails. The tips are always black.

The world's smallest living primates are **Madame Berthe's mouse lemurs**. They weigh just one ounce (28 g). Four will easily fit in a human's hand.

Black-and-white ruffed lemurs are the world's largest pollinators. When they eat flower nectar, they get pollen on their faces that pollinates the next flower they visit.

Sportive lemurs carry beneficial bacteria in their gut. They eat mostly leaves, which are hard to digest. The bacteria help the lemurs break down the leaves.

Red-fronted lemurs chew and crush millipedes to produce a potent paste. When they spread the paste on their skin, it acts as a natural insect repellent.

Indri lemurs sing! Their group singing sessions sound like a pod of whales calling to each other. The sounds play a key role in troop bonding.

Fork-marked lemurs feed almost exclusively on tree gum, which is sometimes called sap or resin. Animals who eat this peculiar diet are called gumnivores.

WORDS to Know

arboreal	living in trees or spending most of the time in trees
digits	divisions of limbs, often in groups of five; fingers or toes
diurnal	active mainly during the day
extinct	not living; died out
locomotion	the act of moving from place to place
nocturnal	active mainly at night
prehensile	able to grip or grasp
primates	mammals that have large brains, hands and feet that grasp, and keen vision; mammals in a group that includes monkeys, apes, and humans
tactile	relating to the sense of touch
vocalizations	sounds that are uttered to communicate emotions
weaned	no longer fed by a mother's milk; able to eat solid food

LEARN MORE

Books

Brydon, Alli. *Go Wild! Lemurs*. Washington, D.C.: National Geographic, 2022.

Jaycox, Jaclyn. *Ring-tailed Lemur Princesses: Rulers of the Troop*. Mankato, Minn.: Pebble Books, 2022.

Kellett, Jenny. *The Ultimate Lemur Book*. Melbourne, Australia: Bellanova Books, 2023.

Websites

"Lemuridae: True Lemurs." Animal Diversity Web. https://animaldiversity.org/accounts/Lemuridae

"Ring-Tailed Lemur." Smithsonian's National Zoo & Conservation Biology Institute. https://nationalzoo.si.edu/animals/ring-tailed-lemur

"The Time Is Now: Lemurs Are the Most Endangered Mammals on Earth." Duke Lemur Center. https://lemur.duke.edu

Documentaries

Douglas, David. *Island of Lemurs: Madagascar*. Burbank, Cali.: Warner Bros, 2014.

Lemurs of Madagascar. Terra Films TV, 2018.

Trouble in Paradise: The Last Lemurs of Madagascar. Free Documentary, Nature, 2023.

Note: Every effort has been made to ensure that any websites listed above were active at the time of publication. However, because of the nature of the Internet, it is impossible to guarantee that these sites will remain active indefinitely or that their contents will not be altered.

Visit

ANDASIBE-MANTADIA NATIONAL PARK

See Madagascar's largest living lemur, the indri, in a park that's a three-hour drive from the capital city.

5F85+2F5
Falierana, Madagascar

DUKE LEMUR CENTER

See over 200 lemurs in the most robust population of lemurs outside Madagascar. Take an in-person tour or adopt a lemur to help support its care.

3705 Erwin Road
Durham, NC 27705

SAINT LOUIS ZOO

Look for exhibits of sifaka lemurs, black-and-white ruffed lemurs, mongoose lemurs, and ring-tailed lemurs.

1 Government Drive
St. Louis, MO 63110

SMITHSONIAN NATIONAL ZOO

Leap onto Lemur Island and catch a training session with ring-tailed lemurs.

3001 Connecticut Ave. NW
Washington, DC 20008

INDEX